Helping Babies Grow

BY KIM THOMPSON

A Little Honey Book

Crabtree Publishing
crabtreebooks.com

Tips for Teachers and Caregivers

This book supports early readers as they decode words to learn facts and gain knowledge about the world.

Before reading, make sure students understand the sound-spelling correspondences shown below as well as the high-frequency words shown on the next page. Introduce the vocabulary words.

During reading, provide feedback and encouragement as students sound out decodable words by blending individual sounds.

After reading, talk about and write about the topic. Share the information on page 16 to help students learn more.

Letters and Sounds

New:

none

Review:

all consonant sounds and standard spellings; consonant digraphs *ch*, *ng*, *sh*, and *th* (voiceless); *short a* spelled *a*, short *e* spelled *e*, *short i* spelled *i*, *short o* spelled *o*, *short u* spelled *u*

Decodable Words

and, at, big, branch, bring, chest, chicks, cling, dad, den, drift, drinking, drop, eggs, fat, fish, flap, flash, fluff, fresh, gets, grab, hang, hatch, help, his, in, is, it, milk, mom, munch, mush, nest, not, on, plants, rich, rush, shell, sloth, snug, spit, spring, stand, swim, thick, trot, twigs, up, which, will, with

High-Frequency Words

New: call, grow, keeps, new, own, show

Review: a, are, does, eat, find, for, from, good, he, make, out, some, soon, the, their, they, to, walk, warm, water, white

Vocabulary Words

babies

elephants

orca

owl

penguin

polar bear

Elephants help their **babies** stand and walk.

They show babies which plants are good to munch.

Owl babies call from the nest.

In a flash, mom and dad owls flap in. They spit up mush for the chicks to eat.

Orca moms swim with their babies. Their milk is thick and rich in fat.

Drinking it helps
babies grow.

Sloth babies cling to their moms' chests. They grab at twigs. Soon, they will hang from a branch on their own.

Polar bear moms find a fresh white drift. They make a den for their babies.

Babies grow big. Their fluff gets thick. They will trot out in the spring.

penguin

Some **penguin** dads keep eggs. The dad does not drop his egg. He keeps the shell snug and warm. It will hatch soon.

Moms rush to the water. They bring fish for the new chick.

Build Background Knowledge

Many animal babies need special care and attention. Parents help by sheltering and protecting their offspring, feeding them, and teaching them survival skills. Babies chirp, call, and cry to communicate their needs. Adults respond by giving food and milk, providing warmth, creating safe places, and much more. What special needs do young humans have? How do parents meet their needs?

Helping Babies Grow

Written by: Kim Thompson
Designed by: Rhea Magaro
Series Development: James Earley
Educational Consultant: Marie Lemke, M.Ed.

Photographs: All images from Shutterstock

Crabtree Publishing

crabtreebooks.com 800-387-7650

Printed in China/012024/FE20231222

Published in Canada
Crabtree Publishing
616 Welland Ave.
St. Catharines, Ontario
L2M 5V6

Published in the United States
Crabtree Publishing
347 Fifth Ave
Suite 1402-145
New York, NY 10016

Library and Archives Canada Cataloguing in Publication
Available at Library and Archives Canada

Library of Congress Cataloging-in-Publication Data
Available at the Library of Congress

Hardcover: 978-1-0398-4442-1
Paperback: 978-1-0398-4523-7
Ebook (pdf): 978-1-0398-4600-5
Epub: 978-1-0398-4670-8
Read-Along: 978-1-0398-4740-8
Audio: 978-1-0398-4810-8